Majed de Lacerda Charafeddine

The influence of ergonomics on employee performance

Majed de Lacerda Charafeddine

The influence of ergonomics on employee performance

An analysis of employee perception of the impact of ergonomics on their own professional performance

ScienciaScripts

Imprint

Any brand names and product names mentioned in this book are subject to trademark, brand or patent protection and are trademarks or registered trademarks of their respective holders. The use of brand names, product names, common names, trade names, product descriptions etc. even without a particular marking in this work is in no way to be construed to mean that such names may be regarded as unrestricted in respect of trademark and brand protection legislation and could thus be used by anyone.

Cover image: www.ingimage.com

This book is a translation from the original published under ISBN 978-613-9-63049-3.

Publisher:
Sciencia Scripts
is a trademark of
Dodo Books Indian Ocean Ltd. and OmniScriptum S.R.L publishing group

120 High Road, East Finchley, London, N2 9ED, United Kingdom
Str. Armeneasca 28/1, office 1, Chisinau MD-2012, Republic of Moldova, Europe
Printed at: see last page
ISBN: 978-620-7-76539-3

SUMMARY

THANKS

I would first like to thank my father, Majed Mohamad Nagib Charafeddine, who has always been by my side, supporting me through difficult times, being a companion by listening to me and a mentor by helping and encouraging me.

Secondly, to my mother, Tatiana de Lacerda Rodrigues, for helping me to walk this path, which I might not be walking today if it weren't for her dedication to helping me throughout my school career.

Thirdly, to my grandmother, who encouraged me to study Production Engineering and who has always been by my side. Last but not least, to all my family members who may have influenced me to complete this stage and have not been mentioned individually.

To my advisor, Professor Janilce dos Santos Negrâo Messias, who gave me total support in carrying out this work, sharing her experience and knowledge and evaluating my work, as well as providing materials that enabled me to move forward with the proposal.

To my girlfriend Bruna de Abreu e Silva, who has always been by my side, enduring the difficult times and sharing the moments of joy. She always helped me when I needed it, with as much commitment as possible.

To the Federal University of Paranà, for having made available the knowledge essential to the realization of this work, during graduation, making this work executable.

To the company that made time available for all the employees, the means to collect the data and allowed the work to be carried out.

SUMMARY

This work aims to study the perception of Ergonomics by the employees of a small chemical industry in Pinhais, Paranà, which is not adjusted in accordance with the regulatory standards and basic precepts of Ergonomics. The methodology used, which is characterized as a case study, involved objective surveys of all the organization's workers, seeking information on performance, productivity and quality of life at work. After this stage, the data obtained was detailed and analyzed in order to understand the correlations and their causes and consequences. Among the conclusions is that high employee turnover is associated with a low level of concern about the ergonomic environment and quality of life at work, leading to a negative perception on the part of the workers in relation to the organization as a whole. This has a direct impact on employee motivation and satisfaction, and an indirect impact on productivity.

Key words: Ergonomics, Productivity, Work environment, Motivation

1 INTRODUCTION

According to Balbinotti (2003), from the 1970s onwards, the environment in which companies existed changed drastically, altering the paradigms in which they found themselves and consequently changing the guidelines with which they acted in the market. Adding value, competitiveness and productivity became the focus, and in order to do so, they had to re-evaluate the way they dealt with the market.

The ergonomic issue that arose in the 1970s, as Balbinotti (2003) points out, led companies to produce and consider even more the value that ergonomics could bring to the economic approach of corporations; by improving working conditions, the impact on workers' health was reduced, leading to lower rates of absenteeism and turnover. Unfortunately, these rates are still high in Brazil, says the author.

In Brazil, according to a study carried out by the Getúlio Vargas Foundation (FGV) in 2014, considering data from 2011, there were around 9 million micro and small companies, corresponding to approximately 27% of the national GDP; in commerce, they are responsible for 53.4% of the GDP and in industry, they represent 22.5% of the GDP. In absolute terms, the corresponding industrial production quadrupled between 2001 and 2011, from R$ 144 billion to R$ 599 billion, in 2011 values. These are significant figures that demonstrate the importance of this type of business for the Brazilian domestic market.

According to Chandler (1990), the most competitively successful companies are those whose founders and top executives have understood the logic of the management enterprise, in other words, the logic behind modern industrial capitalism - the most competitively successful companies are those that manage to motivate their employees by developing their own initiatives, allowing them to take ownership of the production process and improve it in order to increase the efficiency of the process and the effectiveness of adding value to the products in question.

Considering this reasoning, it can be seen that an employee's motivation is directly related to the working environment, together with the management style of the person in charge of the company. Thus, the productivity of this individual will be impacted by the conditions to which he or she is subjected and the management parameters within the company. In other words, the relationship between culture - or organizational climate - and employee motivation has an impact on the success of the company.

1.1 PROBLEMATICS

According to Balbinotti (2003), ergonomics studies the suitability of work for human beings and how to develop it effectively to satisfy the needs of the worker and the company.

According to Gontijo (2001), the ergonomic study of work is aimed at the physical and mental health of the worker, improving productivity through detailed analysis. In other words, the intention of the ergonomic study is to optimize the human resources employed, so as not to saturate them - meaning physical and psychological wear and tear on the worker - and, at the same time, achieve the objectives set by the organization in terms of productivity and quality.

Considering the 82% increase in the turnover rate since 2010, according to Bispo (2013), from Catho, Brazil is the country with the highest rate. The world average, by way of comparison, is 38%. This same survey points to a lack of recognition, low pay, demotivation and concern about the company's future as the main reasons for this figure.

Correlating these figures with the number of accidents at work in recent years in Brazil, according to the TST (Superior Labor Court, 2012), there has been a reduction in the absolute number of accidents, approximately 705,039 occurrences. Compared to 2011, the reduction was approximately 2.2%.

However, as Haje (2014) points out, in a publication on the Chamber of Deputies website and using data from the Ministry of Health, Brazil is the fourth country in terms of the absolute number of deaths from fatal accidents at work. The author also adds that the main causes are the trivialization of occurrences and the lack of prevention policies; she also claims that 10% of GDP, according to the Ministry of Health, is spent on accidents at work - considering all developing countries.

In essence, the subject is a question: what is the view of the employees of a small chemical industry regarding quality of life at work, productivity and motivation, and the impact of management decisions on the above-mentioned bases?

1.2 OVERALL OBJECTIVES

To evaluate the influence of Ergonomics on the quality of work performance of employees in a chemical industry.

1.3 SPECIFIC OBJECTIVES

a.   Checking that the work environment is adapted to the needs of the employees, in line with the precepts of NR-17;

b.   Evaluate employees' perceptions of the organization based on working conditions;

c.   To identify possible illnesses caused by ergonomic problems and acquired by employees as a result of their work;

d.   By means of data collected from employees, we can show that ergonomics, QWL, motivation and satisfaction have an impact on the quality of the work done, the perception of the organization and productivity throughout the working day.

1.4   BACKGROUND

According to the latest survey, there are approximately 8.9 million micro and small businesses in Brazil, accounting for 27% of GDP. They account for 52% of jobs with a formal contract and 40% of salaries paid. It is the largest part of the country's economy (FGV, 2011).

In the company there is an unfavorable scenario for the maintenance and retention of employees in small companies, with high rates of absenteeism and employee turnover, which raise questions related to the motivation and satisfaction of employees with the organization and what are the causes that originated the current scenario; intending to reverse the situation found, on the part of the company, with the data analyzed.

"Concern about ergonomics and its possible improvement contribute directly to improving the efficiency, reliability and quality of organizational operations." GUÉRIN, 2001.

According to Cruz (2012), 85% of Brazilian companies have already implemented initiatives aimed at employee well-being, while 60% are working on structuring more complete programs. He also concludes that large companies with more resources are not always the best in this area.

1.5   METHODOLOGY

This work can be characterized as a case study, mainly because an intense study was carried out on a few objects, in order to allow the collection of broad and detailed knowledge. (GIL, 2008)

According to Yin (2005), a case study is an empirical study that investigates a current phenomenon within its own context of reality, and where several sources of evidence are used.

According to Vergara (2007), research has two basic criteria in terms of purpose, and this work is characterized by exploratory and descriptive research. It was exploratory because it sought out literary sources on the subject, in order to broaden knowledge and improve understanding of the subject, and crossed with data collected in the field, in order to explain the existing problem.

It is characterized as descriptive because it allows the investigation of real facts using questionnaires to solve occupational health problems in the company.

Exploratory research aims to provide greater familiarity with the problem, seeking to make it clearer. Exploratory research is less rigidly planned; it develops, clarifies and modifies concepts and ideas, formulating more precise problems or researchable hypotheses for future studies. They provide an overview of a given fact, and the end product becomes a more clarified problem. (GIL, 2008)

In terms of approach, the research is quantitative and qualitative. Quantitative because it uses a questionnaire, Appendix A, which allows the information obtained to be explored in numbers; the data is treated numerically, considered as blocks that cannot be interpreted outside the context in which they were obtained. This information is the employees' individual assessments of the ergonomic environment in which they work; and it is considered in a macro way, where all the data is presented at once, quantifying the data and generalizing the results.

As for the qualitative approach, it refers to the final part of the survey, in which suggestions for improvements were requested, which must be analyzed by people who have an understanding of the organization's reality. As the number of responses is smaller and cannot be generalized, an interpretative approach is possible, without the need for a statistical survey - as is the case with the quantitative approach.

For quantitative research, the objective answer must be obtained in numerical, unquestionable form; as for qualitative research, the analysis and interpretation of the data must be carried out by the person responsible, and based on this information, a conclusion must be reached. (MICHEL, 2005)

1.5.1 Data collection

One of the main foundations of the work is the collection of initial data, with which it is possible to compare the results obtained during the course of the work.

Over the course of six months, the relevant information was collected, making it possible to correctly size up the problems to be defined.

We used a questionnaire with closed (objective) questions, Appendix A, which was applied between September 1st and 8th, 2014. The questionnaire was unified so that it could be used by all the employees, since the company is small and the functions are shared between everyone; in this way, important parts could be organized in a concise and coherent way with the desired data collection profile. The questions were designed to involve the following different areas:

a.   Evaluation of the perception of physical, cognitive and organizational ergonomics;

b.   Employee health assessment;

c.   Evaluation of the work environment, performance achieved and desired and, finally, any suggestions that might be feasible.

The questionnaire is an instrument for collecting data, made up of ordered questions that must be completed without other people being present. (MARCONI E LAKATOS, 2007)

To make it easier for all the employees to understand, as well as for future analysis of the data, all the questions have five alternatives to answer, with the intention of comparing each of these options with a score, ranging from 1 to 5. In all of them, , a score of 1 corresponds to the worst possible evaluation of the question, while 5 corresponds to the best possible evaluation. Table 1 shows this distribution in schematic form.

TABLE 1 - EVALUATION IN THE QUESTIONNAIRE

Note	Corresponding generic meaning
1	Far below average
2	Below average
3	Average
4	Above average
5	Far above average

SOURCE: the author (2014)

The format of the questionnaire was intended to limit and standardize the answers, making it easier for the respondent to understand and interpret the data.

The questions drawn up are based on establishing a link between the company's problems and objectives, the population to be surveyed, the possible scenarios to be researched and the method of analysis considered (GOODE and HATT, 1972).

As for the advantages and disadvantages, the first are ease of application, process and analysis; ease and speed of response; little possibility of error. The second are preparation time, so that all the questions and possible answers are included; there may be influence by the alternatives presented on the respondent (GOODE and HATT, 1972).

Considering the company's profile, it is believed that this approach was the best way to collect the data, in a satisfactory manner and consistent with the objectives set.

1.5.2 Sources of information

The first stage was to research the scientific literature on the subject, and then go out into the field to gather data. Primary and secondary sources were used, as defined by Marconi and Lakatos (2007).

Field research was carried out with the aim of obtaining data so that it would be possible to analyze the company's current situation, marking out a base line from which the work was carried out.

1.6 CASE STUDY

It is located in the city of Pinhais, in the metropolitan region of Curitiba, in the state of Paranâ. It occupies a constructed area of 625m2.

The company was founded in 2000, a time of expansion in the pet market, with the aim of taking advantage of the high profitability offered by this type of product, with the aim of manufacturing products that were more advanced than the competition.

The workforce has changed three times since the start of the project, during the second half of 2014, to include eight CLT employees and two managing partners. This makes a total of ten full-time employees.

Throughout its existence, it has had three different ownership groups, the last of which took over the company in May 2014.

1.6.1 Market

Called the *pet* market, it corresponds to products and services aimed at pets. This niche emerged in Brazil in the mid-1990s, according to Abinpet (Brazilian Association of the Pet Products Industry).

It is the 2nd largest market in the world, with revenues of R$15.2 billion in 2013 and growth between 2012 and 2013 of 7.3%. For 2014, it is expected to grow by almost 10%, reaching a forecast turnover of R$16.63 billion (ABINPET, 2014).

Within the *pet* market there is a niche called *pet serv*, in which the company finds itself indirectly. It offers products for pet shops, i.e. products that the pet owner consumes secondarily - through the groomer. This specific market grew by 26% in 2013, reaching approximately 19% of total turnover (ABINPET, 2014).

It is estimated that there are approximately 37.1 million dogs and 21.3 million cats in the country. The country has the fourth largest domestic animal population in the world, with 106.2 million individuals (dogs, cats, fish, birds, etc.), and the second largest absolute number of dogs and cats (REZENDE, 2014).

2 THEORETICAL FRAMEWORK

In order to carry out the study, information was needed on concepts, market practices and current legislation, which is explained throughout this work.

2.1 ERGONOMICS

Iida (2005) defines Ergonomics as the study of how work adapts to man, and not the other way around - the difference is clear to see. The author stresses that, in addition to machines and equipment, the situation in which the relationship between man and productive activity takes place should also be considered, involving organizational aspects.

Ergonomics (or Human Factors) is a scientific discipline related to the understanding of the interactions between human beings and other elements or systems, and the application of theories, principles, data and methods to projects in order to optimize human well-being and overall system performance (ABRAHÂO, 2009).

Laville (1977) considers ergonomics to be the body of knowledge associated with man's performance in activity, applying it to tasks, instruments, machines and production systems.

Ergonomists plan and evaluate tasks, workstations, environments and production systems, seeking to make them compatible with the characteristics of each person - needs, abilities and limitations. (ABRAHÂO, 2009)

According to Balbinotti (2003), protecting workers is just one of the aims of ergonomics, since the consequences are improved productivity and quality of work, which are also expected and targeted results.

Wisner (1994) considers that ergonomics is concerned with all the human aspects of work, in whatever situation it is carried out, and has two aims: improving and preserving workers' health, and the design and satisfactory operation of the technical system involved, from the point of view of production and safety.

According to Abrahao (2009), the aim of ergonomics is to transform work in such a way as to adapt it to the characteristics and variability of man and the production process; it develops the project in such a way as to include well-being, safety, productivity and quality in organizations.

2.1.1 History of Ergonomics

Work-related illnesses were first described in the mid-1700s by Ramazzini, an Italian doctor. Its academic origin, however, was in 1857 with an article published by

Jastrzşbowski. Its etymology comes from the Greek words "ergon" and "nomos", which mean work and natural law, respectively. (ABERGO, 2014)

According to Iida (2005), the birth of Ergonomics dates back to July 12th, 1949. On this day, several researchers met to formalize the creation of a new branch of applied science. Although the name emerged at the second meeting of these individuals in 1950, the term Ergonomics first appeared in Jastrzebowski's article, 'Essays on ergonomics or the science of work, based on the objective laws of science about nature' (1857).

Ergonomics was formalized in 1949 by K.F.H. Murrell, who coined the term in world history with the creation of the first national ergonomics association, the Ergonomics Research Society, which brought together professionals interested in adapting work to humans. From this moment on, and as early as 1950, the term was adopted in countries with varying industrialization processes (ABRAHÂO, 2009).

In the 19th century, Frederick Winslow Taylor used the term in his book 'Scientific Management', looking for the best alternatives for carrying out work during the industrial revolution. During the world wars, further back in time, there was an unprecedented cognitive demand, requiring operators to make critical decisions at great speed. Consequently, throughout this period of time, the indirect development of ergonomics was necessary for the competitiveness of industries. (ABERGO, 2014)

According to Abrahao (2009), following the movement after the creation of the Ergonomics Research Society, the HFS (Human Factors Society) and the IES (International Ergonomics Society) emerged in 1959, both in the United States, and in 1963 in France, the SELF (Société d'Ergonomie de Langue Française).

[a]Modern Ergonomics is considered to have emerged at the end of World War II, illustrating the purpose for which it was created. According to Wisner (1994), at the time, the Royal Air Force was trying to understand why the most modern equipment available at the time was not being operated as efficiently and effectively as expected. In order to understand these facts, a team was set up consisting of an engineer, a psychologist and a physiologist. It was then concluded, after several analyses, that it was necessary to organize and standardize the ways in which information was presented for aircraft design, and thus limit reading errors and the possibility of accidents. They thus demonstrated the importance of adapting technological artifacts to the characteristics and limits of our perceptual-cognitive processes (ABRAHÂO, 2009).

Ergonomics was considered a militaristic science and was not taken seriously during the immediate post-war period in the United States. They were often referred to as 'button men', in a satirical way. Although this military connotation still remains in certain forms, the discipline has been used for a variety of non-basic applications (IIDA, 2005).

According to Iida (2005), the International Ergonomics Association was founded, which today brings together the main national ergonomics associations; and the first periodical on ergonomics was *Ergonomics*, published in England since 1957.

According to Abrahâo (2009), the emergence of ergonomics can be defined according to the topics in chronological order:

a.   Jastrzebowski (1857): Essays on Ergonomics;

b.   British Royal Air Force: Accidents;

c.   Ergonomics Research Society (1949);

d.   HFS (1959), IES (1959), SELF (1963);

e.   Ombredane & Faverge: Analysis of Work (1955);

f.   Ergonomics Laboratory, CNAM (1970)

Looking specifically at ergonomics in Brazil, the discipline was first mentioned in the 1960s at USP (University of Sao Paulo), according to ABERGO (2014). The person responsible was Itiro Iida, who developed the first Brazilian thesis in the field, Ergonomia do Manejo. At the same university, Paul Stephaneek introduced the subject to the psychology course (ABERGO, 2014).

At the same time, in Rio de Janeiro, Professor Alberto Mibielli introduced Ergonomics to his medical students at UFRJ and UERJ. The biggest boost was still to come, at COPPE, in the early 1970s, with Professor Itiro Iida and his move towards the Production Engineering program at ESDI/RJ; consequently, indirectly to this event, the first book published in Portuguese focused on Ergonomics appeared.

2.1.2 Physical Ergonomics

One of the most widely used definitions would be ABERGO's own August 2000 definition, considered by the IEA (International Ergonomics Association) to be "related to the characteristics of human anatomy, anthropometry, physiology and biomechanics in their relation to physical activity". Some interesting topics include the study of posture at work, material handling, repetitive movements, work-related musculoskeletal disorders, workstation design, safety and health.

It is understood that the focus on the physical aspects of a work situation corresponds to physical ergonomics; it seeks to adapt the demands that activities make on our bodies to the limits and capacities of the human body, using appropriate interfaces. (VIDAL, 2012)

According to Màsculo and Vidal (2011), physical ergonomics deals with the responses of the human body to physical and psychological loads. It is mainly concerned with the physical aspects of the human-machine interface, involving the anatomical, anthropometric and sensory aspects, with the aim of sizing the workstation, making it easier to visualize important information and manipulate what is essential.

According to Grandjean (1991), companies must follow parameters that enable workers to use their muscular strength more efficiently; this is the context in which Physical Ergonomics focuses.

2.1.3 Cognitive Ergonomics

According to the IEA definition (2000), Cognitive Ergonomics is dedicated to the mental processes that affect the interactions between people and other elements of a system, covering perception, memory, reasoning and motor response.

Some relevant topics for study within the cognitive area of Ergonomics would be mental workload, decision-making, human-computer interaction, stress and training, as long as they involve humans and systems (IEA, 2000).

Cognitive ergonomics was given a boost in the 1980s with the advent of the computer, involving the aspects of how information was fed back to users of a given system. Also known as psychological engineering, it covers mental processes - perception, cognition, attention, motor control and memory - and the consequences on the relationships between human beings and the various elements of a system. (MASCULO E VIDAL, 2011)

According to Guimaraes (2004), Cognitive Ergonomics is the area that "encompasses perceptual, mental and motor processes".

According to Abbad (2004), cognitive ergonomics is concerned in three directions: the cognitive, affective and psychomotor domains. By cognitive, the author means knowledge, comprehension, analysis, synthesis and application; by affective, receptivity, response, appreciation and organization; and by psychomotor, it corresponds to positioning, perception and complete mastery.

2.1.4 Organizational Ergonomics

Organizational Ergonomics concerns socio-technical systems, encompassing organizational structures, policies and processes, with the aim of optimizing the relationship between them (IEA, 2000).

Topics for study also include, according to the organization, communications, group work, participatory design, cooperative work, organizational culture and quality management.

According to Màsculo and Vidal (2011), macroergonomics takes on the role of a new approach to optimizing socio-technical systems, organizational structure, policies and processes; it is directly related to the way work is organized, favoring the socio-technical school of cooperative and participatory work.

According to Pinheiro (2014), organizational ergonomics is macroergonomics, encompassing human resource management, organizational culture, forms of communication and the temporal mode of work. Following his line of reasoning, the advantage of macroergonomics is that it recognizes the work group or team, and not just the individual as a work unit; thus, it is more credible when it comes to considering work in a cooperative way.

Synthesizing the lines of reasoning shown above, organizational ergonomics, also known as macroergonomics, assumes that all work takes place in organizational environments, and starts from this assumption in order to proceed with any analysis. Thus, its aim is to enhance the existing systems in organizations.

2.2 QUALITY OF LIFE AT WORK (QVT)

It is essential, according to Balbinotti (2003), to create suitable conditions for people to develop their creativity and avoid activities that could have a negative impact on their health, going through the contributions of Ergonomics and QWL.

According to Iida (2005), stress occurs when a person receives some kind of stimulus and the body prepares itself psychophysiologically for the response action, but this doesn't happen. Therefore, there is an immediate frustration and the accumulated energy must be dissipated, causing physical and psychological effects that are harmful to health.

According to Màsculo and Vidal (2005), QWL has various meanings, depending on the point of view from which it is evaluated. It can be an emphasis on the tasks performed during professional activity, on the human perception of the work done, or an emphasis on the benefits offered by the company, while others consider the promotion systems within the organization or other strategic organizational factors that are important to the evaluator.

Organizational, ergonomic and psychosocial factors can produce stress reactions in individuals; they can have physiological, psychological and behavioural aspects, such as the 'biomechanical load' of muscles and joints, high levels of catecholamine discharge or adverse states of anima (IIDA, 2005).

The author Balbinotti (2003) emphasizes that the human and social dimension at work must be considered by top management; making people work happily and motivated is a mandatory condition for the survival of organizations.

According to Berro (2007), there are benefits to QWL programs for both employees and companies. For the former, better health, interpersonal relationships, efficiency and concentration, emotional control, among others; for the latter, better productivity, market image, work environment, employee retention and reduced turnover, accidents and health care costs.

2.3 QUALITY, MOTIVATION AND ERGONOMICS

To paraphrase Drucker (1997), the main asset of organizations seeking prosperity is people, and consequently, the priority concern should be with them.

Balbinotti (2003) relates the issues associated with morale and safety to the search for better quality, lower costs and deadlines; he points out that as good morale and a safe environment for people are ensured, the tendency is for the level of quality to increase, impacting on the reduction of costs and deadlines. Consequently, in order to improve these items, it is mandatory to consider working conditions, thus demonstrating a directly proportional relationship.

According to Iida (2005), there is something in human behavior that makes people pursue their goals and objectives, popularly known as 'drive', enthusiasm, impulse or will. The author also gives examples of what he considers to be motivating tasks: setting goals and challenges, receiving periodic information about personal performance or through rewards.

There is a process theory called expectancy-valence, in which people consider expectancy (chance of success) *versus* valence (value of the outcome), and it is considered to be one of the best ways of explaining motivation. (IIDA, 2005)

A content theory would be that of Maslow (1943), in which people are motivated differently for each of the needs: physiological, security, acceptance, ego and self-fulfillment. In this theory, there is a hierarchy corresponding to the order used, in which the person only moves on to the next level if the previous one is satisfied.

According to Balbinotti (2003), in addition to the concern to increase productivity, the investment in technology to control organizational problems shows the concern to remove people from situations where there are risks of accidents, hard work, and which consequently generate dissatisfaction in the people who work there. The consequences of this type of interaction between company and employee have an impact on increasing quality and productivity, with the certainty of a competitive company and a pleasant environment to work in, which in turn translates into quality of life.

According to Juran (1990), quality has always been related to human relationships.

According to Weiss (1991), "people work for rewards. These don't have to be tangible, like money. They can be intangible, as in the case of letting an employee be the leader of a group".

Considering that motivation at work is not simple, an important starting point is understanding the employee's needs. (DAVIS AND NEWSTROM, 1991)

According to Balbinotti (2003), increasing the quality of a product and reducing its costs increases the added value. Therefore, one cannot forget the work factor, and thus that people have a behavioural/motivational component directly linked to the cultural aspects of the organization, which is undeniable for the performance of individuals as professionals.

2.4   ERGONOMIC ANALYSIS OF WORK (AET)

"Improving working conditions can also mean improving production" (MONTMOLLIN, 1990).

According to Guérin (2001), the purpose or aim of AET is to transform work so that the design of work situations does not affect the health of workers, who can exercise their skills, valuing their abilities and achieving the economic goals set by the company.

Ergonomic analysis is essential for the functioning of a company, involving the design of activities, material and organizational resources and even specific training, ensuring the achievement of objectives that include the preservation of the worker's physical, psychological and social conditions (BALBINOTTI, 2003).

AET aims to apply the knowledge of ergonomics to diagnose and correct a real work situation. It was developed by French researchers and is an example of corrective ergonomics. It consists of five stages: demand analysis, task analysis, activity analysis, diagnosis and recommendations (IIDA, 2005).

The starting point of a TEA is considered to be the reflection on a problem, verifying the initial demand, in order to propose ways of intervening in the current situation. The next step is to clarify the context in which the situation is inserted, using an understanding of the technology and organization (BALBINOTTI, 2003).

AET is a study centered on the worker's activity in order to understand problems and propose solutions, valuing real activity and modeling it by integrating the observation of behavior and the understanding of people's conduct (LELLES, 2002).

The evaluation of ergonomics in the workplace is subject to the scrutiny of the workers on site, and their opinion is essential for the successful development of new ergonomic policies. From this perspective, one of the most important criteria for a successful ergonomic project must be the evaluation of the employees/users themselves, who will make use of its benefits on a daily basis.

Following Balbinotti (2003), the construction of knowledge in Ergonomics is based on action, integrating knowledge from different areas:

a.	Workers' view of their own work, working conditions, difficulties, complaints and verbalized problems;

b.	Observation of activities in real work situations and observations of the most important indicators in the situation;

c.	Comparison and analysis of the data, based on information in the literature.

Of course, workers' views are an important source for guiding data collection and accuracy. It does not indicate the inversion of the relationship in scientific research, because this data is used as an initial guide, and the construction of knowledge is based on practice.

2.5 MONOTONY

Monotony is the opposite of motivation. While one stimulates, the other discourages; they are caused by environmental stimuli, which can be motivating or demotivating (IIDA, 2005).

According to Grandjean (1991), monotony is the reaction of an organism to a situation or conditions that are poor in stimuli.

According to Màsculo and Vidal (2005), the exhaustion of the flow of sensory input and the processes of adaptation and habituation (indifference) are the main physiological reasons for the appearance of monotony.

It presents itself as a reaction of the organism to a uniform environment. It therefore occurs in all situations with this characteristic: little stimulation from the environment or little excitement; little satisfaction with work is considered a prerequisite. The symptoms are a feeling of fatigue, drowsiness, sluggishness and decreased vigilance (IIDA, 2005).

There are certain conditions for monotony to occur, including external and personal causes. It can be a dull job, or a person who is tired, or an extrovert who is in a limiting situation; it can also correspond to low satisfaction at work. (GRANDJEAN, 1991)

According to Iida (2005), there are certain conditions that aggravate monotony: short work cycles, short learning periods and restriction of body movements, poorly lit, hot, noisy and socially isolated places.

According to Màsculo and Vidal (2005), it combines external causes with personal conditions, which largely determine the onset of symptoms.

Generally speaking, in operational terms, Iida (2005) says that there are two measurable consequences: a decrease in attention and an increase in response time.

2.6   OCCUPATIONAL HEALTH

Occupational health is an area of health that takes care of workers' health, preventing illnesses or problems caused by work. (SSO - Gestao em Segurança e Saùde Ocupacional, 2014)

It is a requirement imposed by the MTE, aimed at safeguarding the quality of life of workers. (MAXIPAS, 2014)

According to Maxipas (2014), it seeks to provide people with a qualified and healthy environment in which to carry out their work activities.

According to NR 7, the Regulatory Standard that defines the premises of occupational health, the aim is to promote and preserve the health of all workers in a company (MTE, 1978).

2.7   STANDARDIZATION OF ERGONOMICS IN BRAZIL

In Brazil, the NR17 regulatory standard was established in 1990, requiring ergonomics as a basic requirement within companies. The aim of this standardization was to seek improvements and adapt the working environment to people, drastically reducing the incompatibilities found until then between work and employees. Despite having been implemented so many years ago, many micro and small companies still don't follow it consistently, due to a lack of inspection by the Ministry of Labor.

According to Oliveira (2010), the standard was created based on union protests caused by frequent cases of RSI and WMSD, recognizing that these illnesses originated from inconsistent working conditions faced by employees in various professional categories.

All Brazilian labor legislation is based on Decree-Law No. 5,452 of 1943, known as the Consolidation of Labor Laws - CLT. In order to specify what the Law deals with, Regulatory Norms, or NRs, are specified, which detail the issues dealt with in the CLT in a specific and, in its entirety, comprehensive way. There are currently 33 regulatory norms that refer to the working environment, each with a different theme; considering the sum of all of them, the most varied working conditions that may influence the working conditions of Brazilian citizens are mentioned. The specific standard that deals with ergonomics is NR17.

2.7.1 Regulatory Standard NR-17

The Ministry of Labor and Employment's (MTE) definition of Ergonomics for NR-17 (Regulatory Standard - Ergonomics), aims to parameterize working conditions to the psychophysiological characteristics of workers and the nature of the work, providing comfort, safety and performance in the best possible way.

According to Màsculo and Vidal (2011), NR-17 sets out the adaptation of working conditions and the AET; it specifies conditions for the individual lifting, transporting and unloading of materials, for the furniture and equipment of the workstations, for the environmental conditions of the institution and of the organization itself within the company.

2.8   RELATED HEALTH PROBLEMS (LER AND DORT)

Work-related illnesses such as RSI (repetitive strain injury) and WMSD (work-related musculoskeletal disorders) are increasingly in the spotlight in the Brazilian media, aggravating the situation of countless workers in the country because they don't have the right working conditions at their disposal to avoid developing these illnesses. Considering these healthy working conditions as a basic precept for the full development of their functions, workers often fail to achieve the productivity desired by their employers, which causes, in addition to inadequate physical conditions, serious psychological pressures due to the pressure imposed on employees.

A worker's state of health depends on their professional activity, and work-related health problems are not exclusively occupational diseases recognized as such, or accidents at work. Under certain conditions, work has a positive impact on an individual's health (BALBINOTTI, 2003).

Nowadays, there are still very few companies that adopt an ethical approach to employee health and do not only think about productivity, but also assume that the person behind the employee has the right to the preservation of their physical, psychological and social state (CARDOSO, 2012).

These diseases mainly affect the upper limbs, including hands and fingers, making movement difficult, with tingling and muscle fatigue in the affected areas; it can even reduce the range of movement, sensitivity or temperature of the sites.

It should be noted that most of the time, the symptoms are related to inadequate activities not only of the upper limbs; inadequate movements usually involve the whole body, which is susceptible if there is mechanical compression of an anatomical part, or if the person remains in a single position for hours at a time. (VARELA, 2014)

The impact that ergonomics has on the health of employees is shown, and how essential it is for companies to aim for the well-being of people within their facilities.

2.9 PRODUCTIVITY AND COMPETITIVENESS

Productivity can be defined as the proportion of resources invested in a given system, and the results returned. In other words, it is the efficiency of a process. According to the Michaelis dictionary (2014), the definition is "the yield of an activity as a function of time, area, capital, personnel or other factors".

"People influence productivity; they modify productivity; productivity depends on the performance of these people." (BALBINOTTI, 2003)

In industrial work, it has been observed that long working hours reduce employee performance. Even if workers have a forced rhythm, exemplified by machines with constant speeds, it has been observed that errors appear randomly, more and more often. (IIDA, 2005)

Still according to Balbinotti (2003), the ergonomic issue becomes clear when confronted with productivity, since Ergonomics seeks better working conditions, so that it can develop without affecting workers' health. The consequences are a reduction in absenteeism rates in the organization, as well as turnover.

Following the reasoning suggested by Màsculo and Vidal (2005), another factor that substantially impacts the mentality in organizations is the Taylorist culture, which is still present in the business mentality. In this culture, there is a dichotomy of work, where some people think and others do; in this way, there is no use of the knowledge generated between the two groups, and not only that, there is a loss of productivity and knowledge as a result of this lack of integration.

As productivity adds value to the product, in order to reduce its operational costs of manufacture and transport, the difficulty lies in identifying the activities that don't, with the intention of suppressing them. (BALBINOTTI, 2003)

For Porter (1986), the explicit concern is to put into practice competitive strategies that guarantee a sustainable position for organizations.

The challenge for companies to survive and prosper is to seek productivity and added value that is greater than the competition. This competitiveness has as its differential factor the development of initiative in employees, because results are no longer achieved by prescribing tasks. (BALBINOTTI, 2003)

3 ANALYSIS, RESULTS AND DISCUSSION

In subtopics 3.1 and 3.2, without analysis, the data obtained by the employees filling in the questionnaire is shown. Sub-topic 3.3 contains the analysis carried out in the company.

3.1 EMPLOYEE PROFILE

We tried to develop a working methodology that has already been reproduced in other academic articles, in order to gather the relevant information and then analyze the data.

The first part of the survey was directly linked to the personal and professional profile of each employee, seeking to categorize the diversity found within the company. General information such as name, age, education, length of time in the company and sector of work was obtained from management.

Graph 1 shows that 60% of the company's employees are women.

GRAPH 1 - CLASSIFICATION OF EMPLOYEES BY GENDER

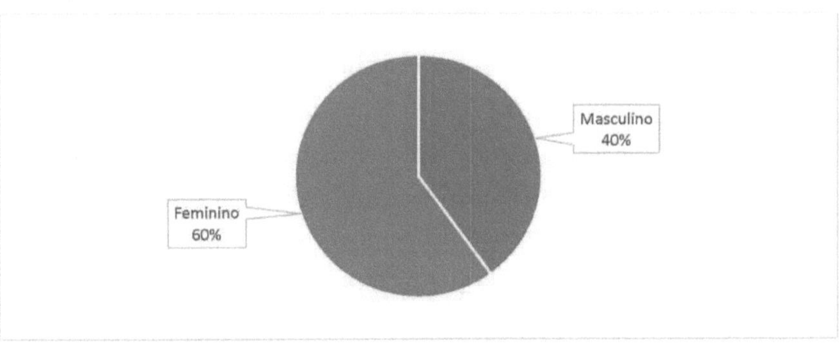

Source: The author (2014)

The management explained that the choice of gender for the employees was aligned with the function they would be performing, i.e. for production, a man is needed because of his physical strength; for labeling and filling, women are preferred because of the need for an exquisite and delicate finish on the products. For the rest of the positions, there are no preferences or limitations.

As well as the choice of employees by age group, which is consistent with the need for work in the face of physical strength, according to the management. As can be seen from Graph 2, most of the company's employees are under the age of 40, corresponding to 70% of the population in question, and 80% are under the age of 49. The average age of the employees is 36.1 years old.

GRAPH 2 - CLASSIFICATION OF EMPLOYEES BY AGE GROUP

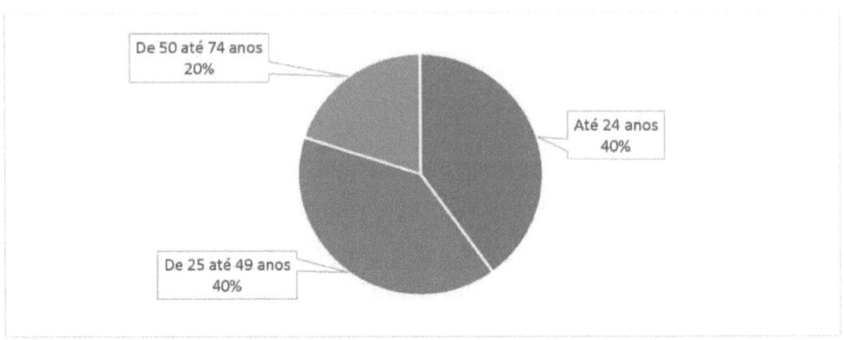

SOURCE: The author (2014)

Still on the subject of age, the manager pointed out the characteristic of having young employees, justifying it by the ratio of work *to* pay, which is not in line with the need for older people. Also according to the management, in order to complement the line of reasoning, the manager disclosed the salary of the employees: the floor of the union of chemical industries in the state of Paranà, which corresponds to R$ 1,036.20 (gross).

Another piece of information highlighted by the company's management is education *versus* age group. Those who are older have less than or equal to high school education, while those who are younger and have completed higher education are recent graduates looking for job opportunities in the labor market.

Graph 3 illustrates the age groups and their proportionality within the company.

GRAPH 3 - CLASSIFICATION OF EMPLOYEES BY EDUCATION LEVEL

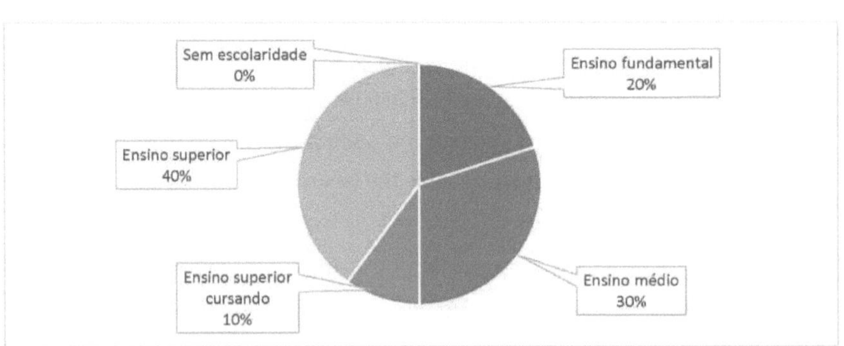

SOURCE: The author (2014)

Graph 4 shows the classification of employees by length of time working for the company.

It is important to note that 90% of the workforce has 7 months or less experience in the company. Another striking fact is that 30% of employees are currently on probation.

GRAPH 4 - CLASSIFICATION OF EMPLOYEES BY LENGTH OF TIME WITH THE COMPANY

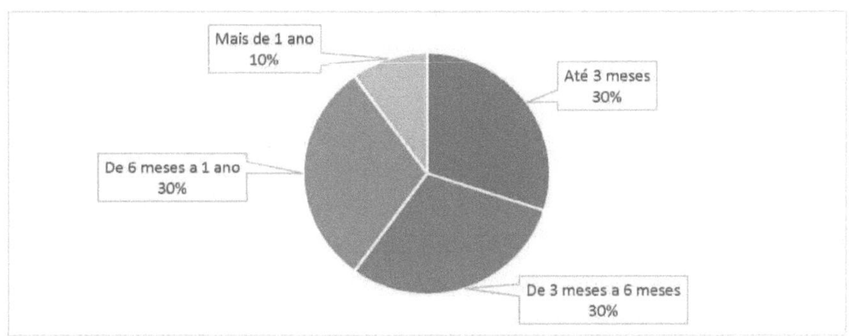

SOURCE: The author (2014)

The organization has a distribution of employees according to the areas in which they work, as shown in Graph 5, in which most of its employees are in two sectors: sales and production. While 20% are in the administrative sector, and allied to this is management, with a further 10% of the workforce. The maintenance sector employs just one person.

GRAPH 5 - CLASSIFICATION OF EMPLOYEES BY SECTOR OF ACTIVITY

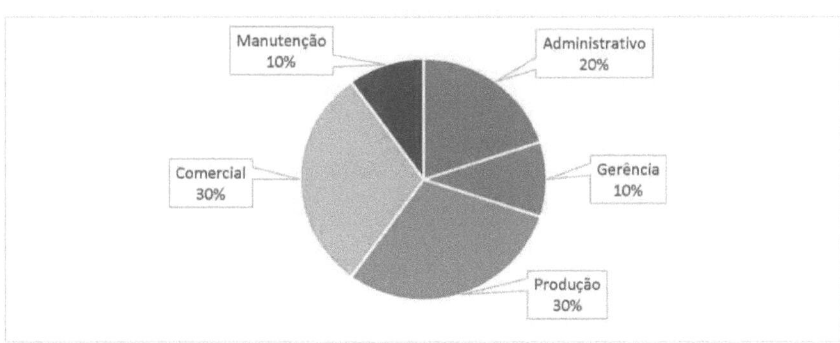

SOURCE: The author (2014)

3.2 WORKING CONDITIONS: ANSWERS TO THE QUESTIONNAIRE

This was the second part of the survey, aimed at evaluating the ergonomic environment from the employees' point of view. It covers more sensitive information, such as the performance achieved, the targets set and the conditions of the working environment.

As stated in subtopic 1.5.1, on data collection, the questions are objective, with answers equivalent to scores between *1* and *5,* with *1* corresponding to well below average, and *5* to well above average.

In each of the graphs, the average was calculated, representing the overall picture in relation to the information provided by the graph. This calculation was carried out using equation 1, which represents the calculation of a simple arithmetic mean:

EQUATION 1 - SIMPLE ARITHMETIC MEAN

$M = (V1 + V2 + V3 + [...] + v_n) / N$

SOURCE: The author (2014)

Where *M* represents the mean itself, *V* are the 'samples' considered, varying up to the maximum number considered - in the company's case, 10, as this is the number of employees - and *N*, which is the total number of 'samples' considered, i.e. the number of *V* considered.

Moving on to the results of the questionnaire, the first question asked about satisfaction with the working day, covering items such as the number of hours per day, workload, targets, pressure and physical effort. The data is shown in Graph 6.

GRAPH 6 - SATISFACTION WITH DAILY WORKING HOURS

SOURCE: The author (2014)

It can be seen that 30% of employees are not satisfied with their working day, considering themselves dissatisfied or very dissatisfied with the environment in which they work; 50% of people are satisfied or very satisfied with their working day, while 20% believe it to be regular, i.e. they are not particularly satisfied or dissatisfied. Considering the data, the average score is 3.2, which means that the overall picture on this issue is one of regularity - not satisfied and not dissatisfied.

With the same aim as the previous question, to assess the working environment in a measurable way, the employees were asked if they considered their work to be monotonous, in question 2. The answers obtained are shown in graph 7. Forty percent of people think that the working day is monotonous, to a greater or lesser degree. Twenty percent consider the routine to be stimulating or very stimulating; and forty percent consider the work to be regular, neither monotonous nor stimulating. The average score is 2.8, meaning that the overall assessment is fair.

GRAPH 7 - MONOTONY IN THE WORK ENVIRONMENT

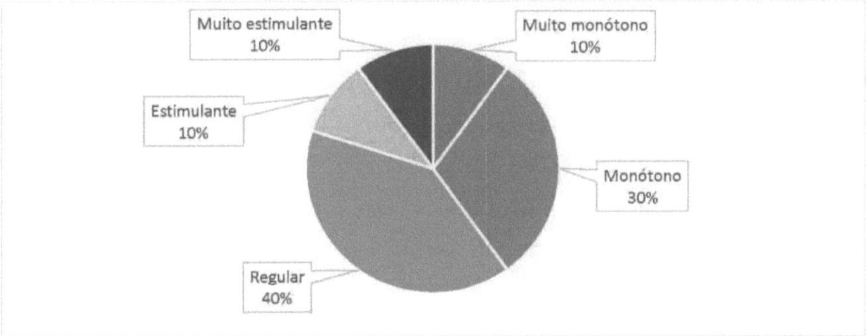

SOURCE: The author (2014)

The next question, number 3, deals with the physical capacity of the employees in relation to the tasks they perform in the company. Only one person - representing ten percent of all workers - stated that the physical effort they usually make is not compatible with their ability.

As for question number 4, whether the activities carried out are compatible with the mental capacity of the employees, all of them answered yes, the tasks are compatible with their psychological conditions.

Regarding the classification of the environment as safe and healthy, which is question number 5, and whose data is shown in graph 8, fifty percent of the employees marked that they did not consider the environment to be safe or healthy - thirty percent considered it to be very unsafe and twenty percent as unsafe. Thirty percent felt that the workplace should be classified as a regular environment, with no tendency towards safety or insecurity. Twenty percent considered the company to be a safe working environment, and none of the employees considered the organization to be a very safe place. The average score on this question is 2.4, between insecure and regular.

GRAPH 8 - PERCEPTION OF SAFETY AND HEALTH IN THE WORKPLACE

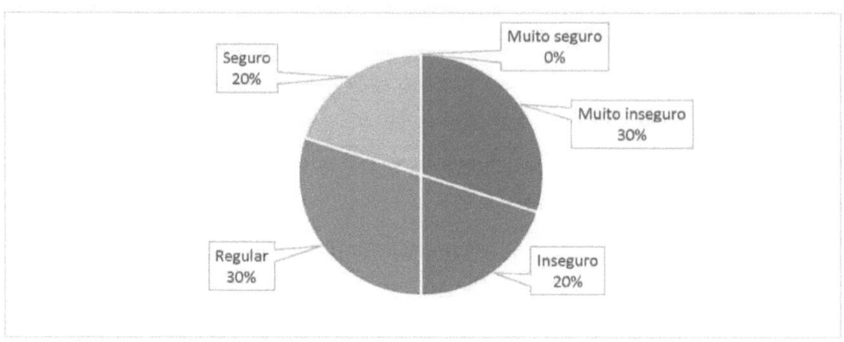

SOURCE: The author (2014)

The workers also answered whether their suggestions are listened to attentively by management and whether their initiatives are encouraged. Graph 9 illustrates the results obtained. Twenty percent of employees consider the environment to be favorable to initiatives, while 30% rate it as fair. Fifty percent consider the environment to be unfavorable to their initiatives. In this regard, the average rating is 2.6, between bad and fair, tending towards fair.

GRAPH 9 - SUPPORT FOR EMPLOYEE INITIATIVES

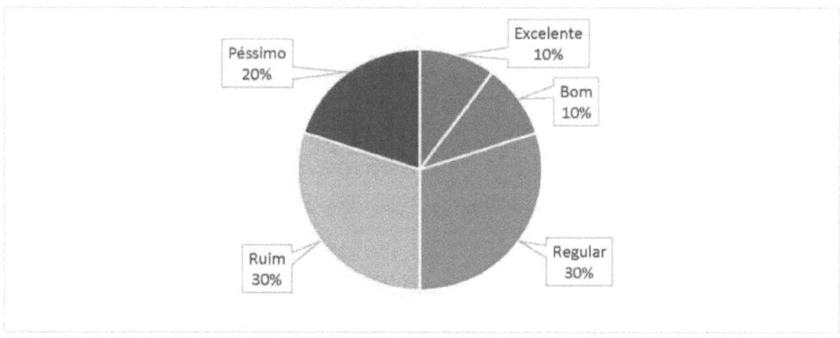

SOURCE: The author (2014)

Question number 7 corresponds to the management's demonstration of concern for the health and well-being of employees. According to the results obtained by , this concern is noticeably felt by only 20% of the people who work at the company. On the other hand, 70% of employees believe that the company does not care about their health and well-being, as shown in graph 10. Ten percent consider the environment to be fair. The average score in this evaluation is 2.1, which is equivalent to a poor overall picture.

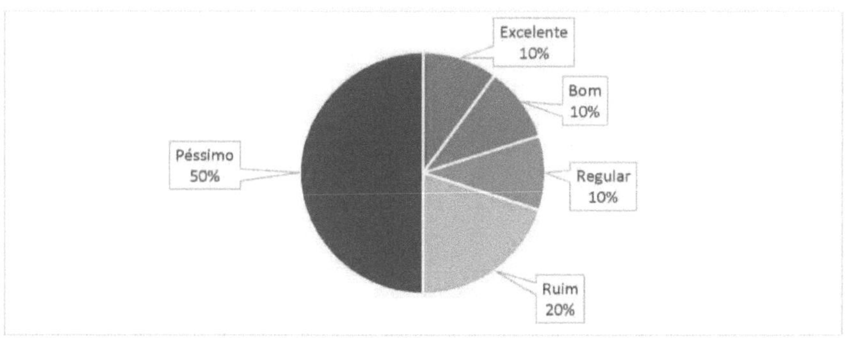

SOURCE: The author (2014)

Questions 8 and 9 refer to the physical and psychological symptoms that employees experience at the end of a working day, if any and what they are. The data obtained was compiled into two graphs, Graph 11 and Graph 12.

As you can see, ninety percent of the company's employees have symptoms after a day's work, to a greater or lesser degree.

As for question number 9, referring to graph 12, seventy percent of the workers have back pain, while sixty percent have joint pain and fifty percent have upper limb pain. Forty percent have neck pain and psychological changes of some kind, including stress or irritability, for example.

GRAPH 11 - PHYSICAL AND PSYCHOLOGICAL SYMPTOMS OF THE DAILY WORK ROUTINE

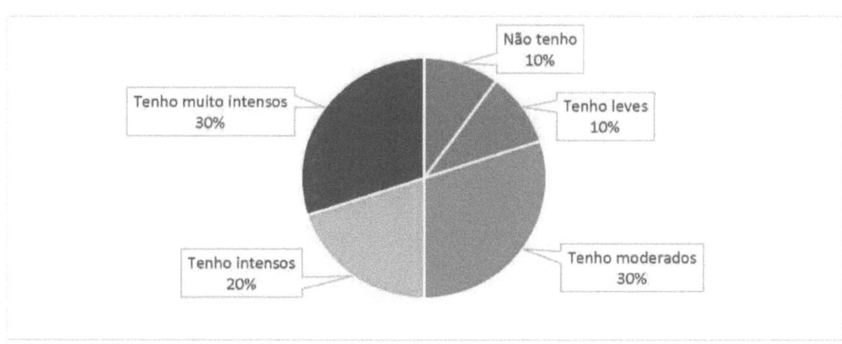

SOURCE: The author (2014)

CHART 12 - PHYSICAL AND PSYCHOLOGICAL SYMPTOMS OF THE WORK RUTINE

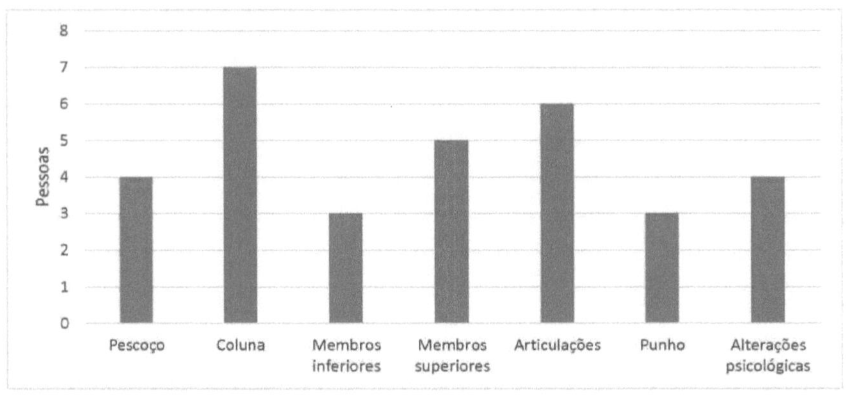

SOURCE: The author (2014)

The average score on the data in graph 11 is 2.5, with the overall picture tending towards moderate symptoms.

The next question, number 10, asked about PPE (personal protective equipment) and appropriate work tools. The result, as shown in graph 13, was that 50% of employees do not consider the supply to be adequate, falling short of what they consider necessary for the company. Thirty percent consider the supply to be equivalent to the standard practiced within organizations, i.e. within the normal range; and 20% consider it to be good or excellent. The average score in this area is 2.6, meaning that the overall assessment is poor, tending towards regular.

GRAPH 13 - SUPPLY OF SUITABLE EQUIPMENT AND TOOLS

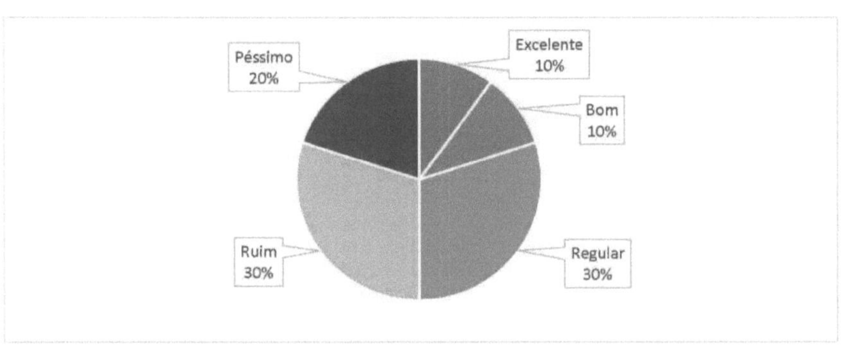

SOURCE: The author (2014)

Question 11 dealt with breaks during working hours, with a focus on relaxation and relieving tension. The answer was that there is an afternoon snack break, and although it is not intended for the physical recovery of workers, it can be used for this. The unanimous response, however, was that there are no exercises for this purpose, relaxation and tension relief.

When asked about the general assessment of the ergonomic environment in the company, question 12, and how this impacts on each person's individual performance, question 13, graphs 14 and 15 were obtained. Fifty percent consider the environment to be irregular and 20% consider it to be positive. Thirty percent consider the place to be regular, i.e. within the normal range. The average score for this question is 2.4, giving the environment a bad rating.

GRAPH 14 - ERGONOMIC ASSESSMENT OF THE WORK ENVIRONMENT

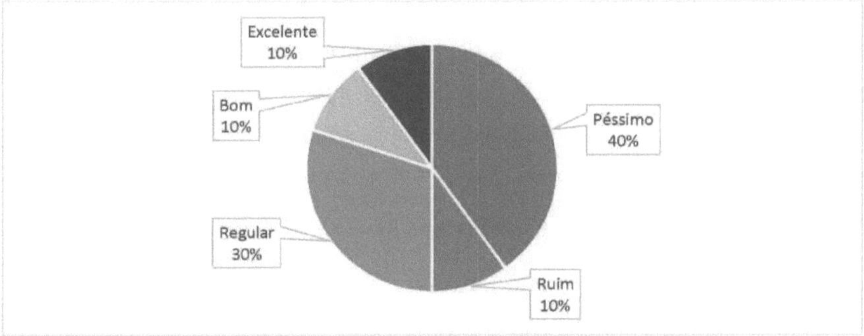

SOURCE: The author (2014)

With regard to individual performance being affected by ergonomics in the workplace, 50% consider it to be affected or very affected and 10% consider it to be little affected. Ten percent consider themselves to be within a hypothetical normal range - in other words, they don't believe they are affected by ergonomics in the workplace. And 30% say their performance is not affected by the ergonomic conditions. The data is shown in Graph 15, and the average score is 2.8, i.e. they consider their performance to be little affected by ergonomics.

GRAPH 15 - PERFORMANCE IN THE ERGONOMIC ENVIRONMENT

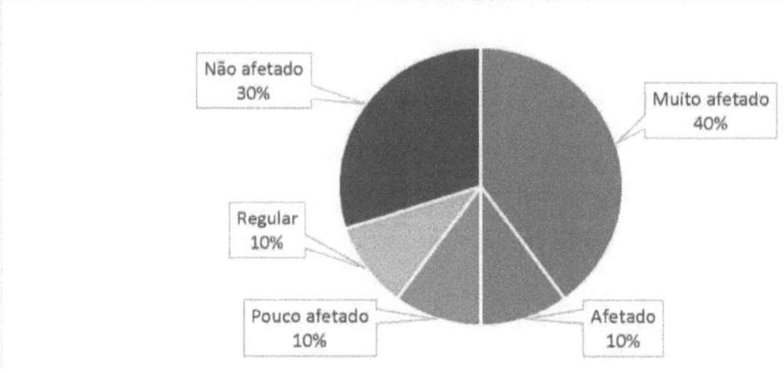

SOURCE: The author (2014)

To conclude the questionnaire, we asked about interpersonal relationships within the company between all the employees, and the results are shown in graph 16. Again, the data is significant, as 40% of employees consider their relationships with colleagues to be poor. Thirty percent consider it fair, and another 30% consider it good or excellent. The average score is 3.1, which corresponds to a fair overall picture.

GRAPH 16 - INTERPERSONAL RELATIONSHIPS IN THE WORKPLACE

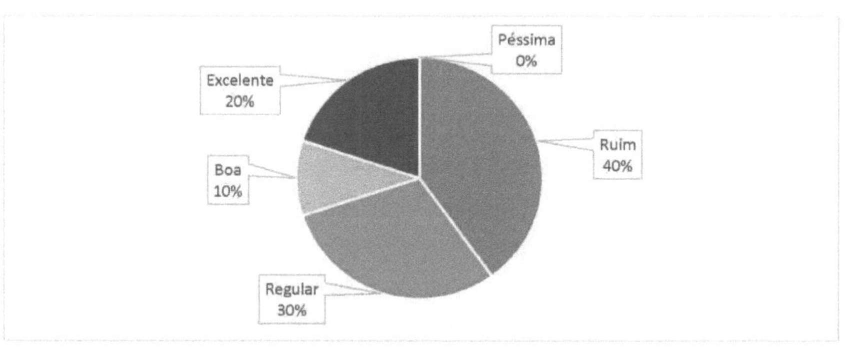

SOURCE: The author (2014)

At the end of the questionnaire, the employees were asked for suggestions for improvements in the development of the company as an ergonomic and social environment, with the aim of improving the scenario surveyed. Five people responded with their ideas for developing the work carried out, which are listed below:

- Reorganizing the internal layout of raw materials and finished products;

- Adjust the lighting in the room;

- Organizing the company's areas;

- Remove items that are no longer used from stock;

- Adapt workstations to the individual needs of each employee;

- Create a weekly meeting for everyone to exchange information;

- Create a daily activity log.

3.3 GENERAL EVALUATION

When comparing the survey data to determine the profile of the employees and the working conditions offered by the company, interesting delineations emerged linking causes and consequences.

Graph 4 shows that the percentage of employees with less than 7 months in the company corresponds to 90% of the total; and also that 30% are in a trial period, not being sufficiently experienced in their respective positions.

Graphs 4 and 6, on the other hand, show a high turnover of employees, resulting in constant renewal and depending a lot on the ability of new employees to learn the job quickly. As a result, people wear out.

As a result, the company is unable to evolve its internal procedures organically and naturally, and often needs external help through consultancies. This is because employees, once they have acquired the practice and experience to suggest improvements and improve operations, end up leaving the company. And the company doesn't take advantage of the knowledge generated while the employee was working there.

Considering the data obtained, the company demonstrates that it does not have a strategy for retaining employees or taking advantage of the knowledge generated; it was later confirmed by management that this is not a relevant issue for the organization at the moment. This makes it difficult to sustainably develop a culture focused on results, with a lack of long-term commitment on the part of everyone involved, both employees and the company.

It's worth noting that half of the employees don't feel comfortable sharing the knowledge they've acquired about the work they do and are free to make changes to internal processes in order to optimize tasks. This influences the results achieved by the company, as it discourages employees from looking for solutions that can improve the production process in any way - reducing time and the number of defects, for example.

Combining the results of the graphs, it is clear that the employees are highly dissatisfied with the company and the situations they find themselves in. After analyzing the responses, it is clear, and the feedback from management confirms this, that there is no concern about retaining employees or the value they can add to operational processes.

To improve working conditions for employees, a series of measures need to be taken. Initially, a QWL program needs to be set up, seeking immediate solutions to the physical overload to which some employees are subjected.

Through these initial procedures, we are looking for the trust and motivation that employees must have for the company in which they work, as this will be the differential in achieving the competitiveness that is currently plaguing the least prepared companies on the market; these are the strategic issues that have a direct impact on productivity, individual performance and the motivation of each employee.

According to Table 2, which summarizes the average indices for each of the questions asked, it can be seen that the overall picture of company-employee satisfaction is low, at 2.65 - which mathematically responds to regularity, although the employee's perception is below average, i.e. negative. To arrive at this final figure, equation 1 was used to calculate the simple arithmetic mean.

TABLE 2 - SATISFACTION RATES

Theme	average satisfaction index
Satisfaction with working hours	3,2
Monotony	2,8
Safety and health	2,4
Support for initiatives	2,6
Concern for health and well-being	2,1
Physical and psychological symptoms	2,5
Suitable equipment and tools	2,6
Ergonomic assessment of the environment	2,4
Performance through the ergonomic environment	2,8
Interpersonal relationships at work	3,1
Overall Average	**2,65**

SOURCE: The author (2014)

The data shown in graph 16 helps to clarify a little more the reasons found for the reality revealed by the survey and outlined in the previous paragraphs, in which 40% of employees label interpersonal relationships with colleagues as poor. This has a direct impact on everyone's perception of the work environment and the company as a whole, and affects the performance achieved as a group of professionals.

3.3.1 Regarding NR-17

There are non-conformities between the Regulatory Standard and the company's practices:

a.   17.2.2 and 17.2.3: manual transportation of loads is carried out without the appropriate technical means, with a single worker supporting more than the appropriate weight (200kg) in metal drums with flammable contents, without adequate training;

b.   17.3.2: the furniture at the workstations is inadequate, has no adjustment and prevents employees from sitting down, so they have to stand for more than four hours in one position;

c.   17.3.3: one of the workers does not have a proper seat, without adequate height adjustment, stressing the popliteus muscle;

d.   17.5.2: effective temperature index above 30°C (maximum allowed 23°C) and relative humidity below 30% (minimum allowed 40°C);

e.   17.5.3: lack of adequate lighting during working hours;

f.   17.6.3: the company does not provide rest breaks, even if there is muscle overload in the neck, shoulders and upper limbs;

g.   17.6.4: the breaks (10 minutes every 50 minutes of work) regulated by the standard are not taken;

These variations from the norm were noted over a period of 6 months, the duration of the work, in several occurrences, and were characterized as normal within that working environment. They were not properly recorded by management.

3.3.2 Physical ergonomics

Along these lines, we found that the organization's Physical Ergonomics is out of balance with current standards and basic conditions that seek to guarantee the health of workers. The furniture is worn out from years of use and was not designed for the purpose to which it is subjected on a daily basis. The tools used to handle the materials are unsuitable for a chemical company, which should follow health surveillance regulations.

Inadequate tooling is characterized by funnels made of PET plastic instead of stainless steel; inadequate storage conditions for products, packaging and raw materials; the manufacture of products without the correct diligence when weighing and mixing the components, where employees are subjected to inadequate physical posture, among others.

3.3.3 Cognitive ergonomics

In terms of Cognitive Ergonomics, information is not made available according to the need to meet stipulated targets, in other words, employees depend exclusively on management to obtain it, and there is no planning, even on a weekly basis, to achieve minimum productivity. Employees work in a kind of "information penumbra", or in other words, they receive information to deal with each time interval, shorter than a shift, as the management deems necessary. There is no sharing of information between employees and the company. As employee turnover is high, an exchange of experiences could be considered, even with management, with the aim of developing an internal culture of knowledge transfer.

3.3.4 Organizational ergonomics

When it comes to Organizational Ergonomics, on the other hand, the company must give employees more credibility to master routine activities, in order to improve productivity and increase satisfaction levels. It should be borne in mind that employees are people who work with the needs of the organization on a daily basis and have knowledge that is often not passed on.

4 FINAL CONSIDERATIONS

Ergonomics focuses on the human aspect of institutions, helping to improve people's performance in terms of productivity and quality of life. The aim is for employees to feel satisfied with their work, without compromising their physical and psychological well-being.

Based on this approach, the research proposal was achieved. To this end, data was initially collected so that the company's current situation could be analyzed in line with daily practices, and not with a fictitious approach to what should be done. To this end, the company lived up to our expectations.

The survey and analysis carried out allowed the company to situate itself within the current reality of the perception of quality of life and satisfaction with the organization, in order to re-evaluate itself and conclude what measures to take in order to improve working conditions. It also had an impact on the interviewees themselves, making them reflect on what improvements could be put into practice, and demonstrating that the company is open to improvements, even if their perceptions were different.

The work generated a positive discussion in order to improve the current conditions imposed on employees. Ergonomic projects, which until then had not been structured to optimize the work routine, have been included in the investment projections for the coming months.

During the course of the study, it was noted that the majority of employees had worrying symptoms, which could easily develop into more advanced clinical conditions and could lead to future occupational illnesses. This problem could be drastically reduced if the institution focused part of its efforts on minimizing the impact on health by investing in ergonomic projects and replacing unsuitable furniture and equipment.

Based on the data collected, it can be seen that employees rate their working conditions as precarious and unsatisfactory, emphasizing ergonomics as essential within a company. To this end, there is a lack of management commitment to achieve the goal of improving the reality for workers. Changes with low operating costs would be of great value in reducing the impact, at least in the initial phase, of the work routine.

The research shows that employee well-being is essential for individual performance with a focus on productivity. It is therefore important for there to be cooperation between the company and its employees, so that the resulting context is beneficial to both.

4.1 SUGGESTIONS FOR FUTURE WORK

It is hoped that the knowledge generated can be used in the future in other related research, including the application of ergonomic assessment methods in order to achieve substantial improvements within other organizations with limited financial resources.

We would also recommend further work focused on implementing ergonomic projects in the company mentioned, with a view to developing ergonomic policies that focus on quality of life associated with the individual performance of each worker. Subsequently, it may be interesting to follow up and collect data again, with the aim of comparing different realities within the same organization, and how concern for ergonomics can be a factor in competitiveness and differentiation between micro and small companies.

5   BIBLIOGRAPHICAL REFERENCES

ABBAD, Gardènia da Silva; BORGES-ANDRADE, Jairo Eduardo. **Human learning in work organizations.** In: ZANELLI, J. C.; BORGES-ANDRADE, J. E.; BASTOS, A. V. B. (Orgs.). Psychology, organizations and work in Brazil. Porto Alegre: Artmed, 2004.

ABRAHÂO, Jùlia. SZNELWAR, Laerte. SILVINO, Alexandre. SARMET, Mauricio. PINHO, Diana. **Introduction to Ergonomics - from practice to theory.** 1ª . ed. Sao Paulo: Edgard Blucher LTDA, 2009.

ABERGO. **Ergonomics - Concepts, Origins, Chronology.** Available at: <http://www.ergonomia.com.br/htm/historico.htm> Accessed on: 10/09/2014.

ABINPET. **Abinpet releases consolidated pet market data for 2013.**

Sao Paulo. Available at: <http://abinpet.org.br/imprensa/noticias/abinpet-divulgadados-mercado-pet-2013/> Accessed on: 20/08/2014.

BALBINOTTI, Giles. **Ergonomics as Principle and Practice in Companies.** 1st ed. Paranà: Genesis, 2003.

BERRO, Diego. **Quality of Life and Productivity.** My article. 2007. Accessed on: 01.11.2014.< http://meuartigo.brasilescola.com/economia-financas/qualidade-vida-productivity.htm>

BISPO, Nathaly. **Brazil has the highest turnover rate.** Available at: <http://www.catho.com.br/carreira-sucesso/gestao-rh/brasil-tem-o-maior-indice-de-turnover> Accessed on: 16/10/2014

BRAZIL, Ministry of Labor and Employment. **10.3 Ergonomics - physical space, equipment, work organization** Available at: <http://portal.mte.gov.br/fisca_trab/10-3-ergonomia-espaco-fisico-equipamentos-organizacao-do-trabalho.htm> Accessed on: 01/11/2014

BRAZIL, MTE, Safe Work. **National Data.** Available at: <http://www.tst.jus.br/web/trabalhoseguro/dados-nacionais> Accessed on: 14/10/2014

CARDOSO, Bianca da Silva. **Prevalence of Musculoskeletal Disorders in Bank Workers in the Campos Novos Region.** 2012. Available at: <http://www.fisioweb.com.br/portal/artigos/categorias/47-art-reumatologia/1325-prevalência-disturbios-osteomusculares-em-bancarios-da-regiao-de-campos- novos.html>. Accessed on: September 21, 2014.

CHANDLER JR., Alfred. **Scale and Scope: The Dynamics of Industrial Capitalism.** Cambridge, Mass.: The Belknap Press of Harvard University Press, 1990.

DAVIS, K. and NEWSTROM, J. W. Human behavior at work - A psychological approach. Sao Paulo: Pioneira, 1992.

DRUCKER, Peter. **The Global Economy and the Nation-State.** Foreign Affairs, Vol. 76, no. 5 (1997).

FGV/ SEBRAE. Micro and small companies generate 27% of Brazil's GDP. Available at:<http://www.sebrae.com.br/sites/PortalSebrae/ufs/mt/noticias/Micro-e-pequenas-

empresas-geram-27%25-do-PIB-do-Brasil> Accessed on: 25/09/2014

GIL, Antonio Carlos. **Methods and Techniques of Social Research.** 6ª . ed. Sao Paulo: Atlas, 2008.

GONTIJO, Leila, et al. Ergonomics applied to the study of work-related musculoskeletal disorders in computers. Anais. ABERGO, 2001. Gramado, RS, 2001

GOODE, Willian J.; HATT, Paul K. **Métodos em Pesquisa Social.** 4th ed. Sao Paulo: Nacional, 1972.

GRANDJEAN, Etienne. **Ergonomics Manual - Adapting work to man.** 4th ed. Rio Grande do Sul: Artes Médicas Sul LTDA, 1991.

GÜÉRIN, F. et al. Understanding work in order to transform it: the practice of ergonomics. Sao Paulo: Edgard Blucher, 2001.

GUIMARAES, Lia B. de M. **Cognitive Ergonomics. Product and Production**. Porto Alegre, 2004.

HAJE, Lara. **Brazil is the fourth country in the world for the number of fatal accidents at work.** Edition: Cronemberger, Daniella. Brasilia, 2014. Available at:

<http://www2.camara.leg.br/camaranoticias/noticias/TRABALHO-E-

WELFARE/471140-BRAZIL-AND-THE-FOURTH-COUNTRY-IN-THE-WORLD-IN-NUMBER-OF

FATAL ACCIDENTS AT WORK.html>

IIDA, Itiro. **Ergonomics - Design and Production.** 2nd ed. Sao Paulo: Blucher, 2005.

JURAN, Joseph M. Juran **Planning for quality.** Sao Paulo: Pioneira, 1990.

LAKATOS, E. M. & MARCONI, M. A. **Metodologia do trabalho cientifico.** 7th ed. Sao

Paulo: Atlas, 2007.

LAVILLE, Antoine. **Ergonomics.** Sao Paulo: Ed. Pedagògica e Universitària Ltda. / EDUSP, 1977.

LELLES, Sérgio Luis Camillo de. Ergonomics and working conditions The case of an oil factory. 2002.

LIMA, Leila Bonfietti. **Brazil: 2nd largest pet market in the world. How are you doing?** Campinas, 2013. Available at :

<http://www.revistapetcenter.com.br/materias/ler-materia/69/brasil-2-maior-mercado- pet-do-mundo-como-voce-esta-nessa> Accessed on: 18/08/2014.

MASCULO, Francisco Soares (Org.). VIDAL, Mario Cesar (Org.). **Ergonomics: Proper and efficient work.** 1st ed. Rio de Janeiro: Elsevier/ABEPRO, 2011.

MASLOW, A. H. **A theory of human motivation.** Psychology Review, [s.l.], 1943.

MAXIPAS. **O** **what is occupational health** . Available at:

.http://www.maxipas.com.br/principal/home/?sistema=conteudos|conteudo&id_conteudo=380> Accessed on: 08/08/2014

MAYO, Elton. The Human Problems of an Industrial Civilization. Taylor & Francis Publishing, 2003

MICHAELIS. **Online Portuguese Dictionary** . Available at:

<http://michaelis.uol.com.br/moderno/portugues/index.php?lingua=portugues-portugues&palavra=produtividade> Accessed on: 09/10/2014

MICHEL, Maria Helena. Methodology and scientific research in the social sciences. Sâo Paulo: Atlas, 2005.

MONTMOLLIN, M. **Ergonomics.** Lisbon: Instituto Piaget, 1990.

OLIVEIRA, Rogéria Bernardo de; OLIVEIRA, Margarete T. Fabbris de. Fabbris de. **Beneficios da Ergonomia fisica, cognitiva e organizacional para as empresas. 2010.** Available at: <http://www.administradores.com .br/artigos/econom ia-e-finanças/beneficios-da- ergonomiafisica-cognitiva-e-organizacional-para-as-empresas/48442/>. Accessed on: October 27, 2014.

PINHEIRO, Monica. **Organizational Ergonomics.** Available at: <

http://www.sistemaambiente.net/monica_pinheiro/monica_pinheiro_ergonomia_organiz acional.htm> Accessed on: 10/11/2014.

PORTER, Michael E., **Competition in Global Industries.** Harvard, Harvard University Press. 1986.

REZENDE, Jàder. **Pet market advances rapidly and will grow by almost 10% in 2014.** Belo Horizonte, 2014. Available at: <http://www.otempo.com.br/capa/economia/mercado-pet-avan%C3%A7a-r%C3%A1 pido-e-crescer%C3%A1 -quase-10-em-2014-1.775966> Accessed on: 22/10/2014.

OHS (Occupational Health and Safety Management). **What *is* Occupational Health?** Available at: <http://www.saudeocupacionalsp.com.br/o-que-e-saude-occupacional.html>Accessed: 13/09/2014

VERGARA, Sylvia Constant. **Research projects and reports in administration.** 8. ed. - sâo Paulo: Atlas, 2007.

VIDAL, Mario Cesar. **Introduction to Ergonomics.** Available at: <
http://www.ergonomia.ufpr.br/Introducao%20a%20Ergonomia%20Vidal%20CESERG.p df> Accessed on: 25/09/2013.

WEISS, D. Motivation and results - How to get the best out of your team. São Paulo: Nobel, 1991.

WISNER, A. A inteligência do trabalho: textos selecionados de Ergonomia. Sâo Paulo: Fundacentro, 1994

YIN, Robert K. **Case study: planning and methods**. 3 ed. Porto Alegre: Bookman, 2005. Classic work

APPENDIX A - QUESTIONNAIRE

Dear contributor, this document is part of an academic project to complete a degree in Production Engineering at the Federal University of Paraná (UFPR), and it would be very interesting if you could contribute to it so that the proposal has a positive outcome. It is not possible to identify the document, i.e. you can answer it without worrying about the answers you provide; they will be used exclusively for the research carried out.

Instructions

Please answer using only one of the alternatives for each of the questions. If your choice does not fit the alternatives, please write a comment next to the question.

How satisfied are you with your working hours?

a.   Very dissatisfied

b.   Dissatisfied

c.   Not satisfied and not dissatisfied

d.   Satisfied

e.   Very satisfied

You consider work:

a.   Very monotonous

b.   Monotone

c.   Regular

d.   Stimulating

e.   Very stimulating

What is your perception of safety and health in the workplace?

a.   Very insecure

b.   Insecure

c.   Regular

d.   Insurance

e.   Very safe

How do you rate the support for your initiatives?

a.   Very bad

b.   Bad

c.   Regular

d.   Good

e.   Excellent

How do you feel about the company's concern for your health and well-being?

a.   Bad

b.   Bad

c.   Regular

d.   Good

e.   Excellent

Do you experience physical and psychological symptoms after a day's work?

a.   Yes,   I have to be   very   intense

b.   Yes,   I have   intense

c.   Yes,   I   have

d.   Yes,   I have   light

e.   I don   't

Where do you feel the symptoms? NOTE: You can mark more than one alternative on this question only.

a.   Neck

b.   Column

c.   Lower limbs

d.   Upper limbs

e.   Articulations

f.   Fist

g.   Psychological changes

How do you rate the provision of adequate equipment and tools for the job?

a. Bad

b. Bad

c. Regular

d. Good

e. Excellent

How do you rate the ergonomics of your work environment?

a. Bad

b. Bad

c. Regular

d. Good

e. Excellent

Do you think your work performance is affected by the ergonomics of your environment?

a. Not affected

b. Regular

c. Little affected

d. Affected

e. Very affected

How do you feel about interpersonal relationships in the company?

a. Bad

b. Bad

c. Regular

d. Good

e. Excellent

yes
I want morebooks!

Buy your books fast and straightforward online - at one of world's fastest growing online book stores! Environmentally sound due to Print-on-Demand technologies.

Buy your books online at
www.morebooks.shop

Kaufen Sie Ihre Bücher schnell und unkompliziert online – auf einer der am schnellsten wachsenden Buchhandelsplattformen weltweit! Dank Print-On-Demand umwelt- und ressourcenschonend produzi ert.

Bücher schneller online kaufen
www.morebooks.shop

Printed by Books on Demand GmbH, Norderstedt / Germany